PLAN ON A JOB

Elizabeth A. Sullivan   2015

We live for future technologies, all those new items and gadgets that excite us. From cell phones to internet, to new apps, we thrive on the best, quickest, and most current information that we can obtain for our daily activities. Today's youth seems to rely on accuracy and being in the know!

Having knowledge of all these technologies is definitely advantageous, but alone will 'not pay the bills.' However, with these technologies we can create our jobs for future generations.

What job might you create?

↕

↕

↕

What about creating jobs applying much of the technology already used on a day to day basis?

What about jobs using 'drones' (unmanned aerial systems)?

Imagine an organization that perhaps does not exist today, but may in the future. What would it be?

I'll create one...

WHAT

IS

IN

YOUR

FUTURE?

The name of my organization is /Person Space (that is pronounced "forward slash Personal Space") and its purpose is to expand our horizons and efforts in exploring new venues for discovering the humanitarian need for drones and using their simple application for reaching necessary goals.

/PERSONAL SPACE

Expand Horizons

Limit suffrage of Human Race

Opportunity to Career Path

Locations Close to Home

Work with Federal Aviation Administration

Learn and Apply Current Technologies

Work Independently

Of course, this company is only a plan for developing in the future, at this point, but imagine an organization that hopes to provide safe operation of unmanned aerial systems, or drones, and will provide employee benefits, as previously mentioned. At the same time, /PS drones could be applicable to human needs such as pursuing cars that flee accidents, checking pole lines after weather-related conditions have left them disarray, or even providing immediate visual surveillance to those roller coaster riders who have become stranded atop their ride. While doing so, otherwise hazardous conditions are kept in check for police/fire, and emergency crew personnel.

✖ UAS: Unmanned aerial system

How might one advertise their job of the future?

Expanding our horizons, */Personal Space* is coordinating efforts with the Federal Aviation Administration in recruiting individuals who are willing to learn new skills, and be challenged with the birthing of a job frontier!   Rooted here in Akron, Ohio, our company is seeking college graduates, or the experienced equivalent, who are ambitious to discover the humanitarian need for drones and their use in simple application for reaching necessary goals. Full-time and part-time work is offered. The ElecMech & Operator Specialist will earn from 90K +, as commensurate with your knowledge, skills, and abilities. Applications are accepted at our company address www./psflight online, or call our office to complete your application process at
330-999-8888.

After researching your job suggestion, presenting *appropriate* information will be necessary.

(/PS research: *These unmanned aerial systems have a history as far back as 1849, when the Habsburg Austrian Empire launched 200 pilotless balloons armed with bombs against the revolution-minded citizens of Venice. They were also used by Union Confederate Forces, to provide intelligence.)

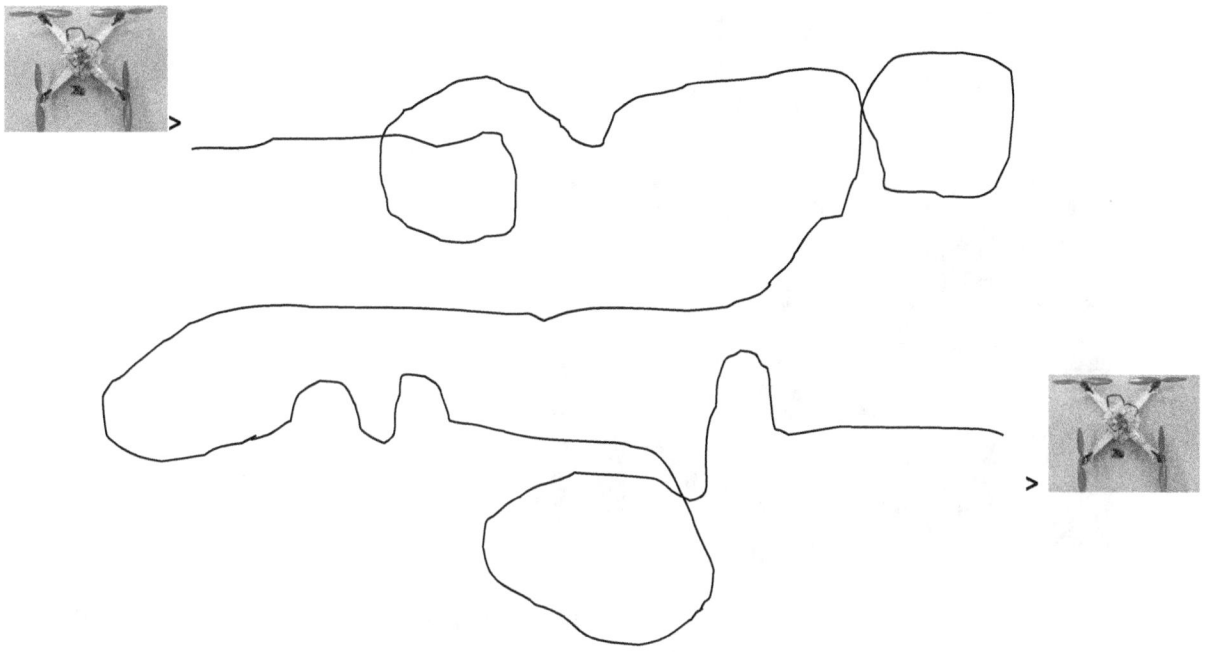

Civil / domestic use of drones

*The history for (commercial) civil or domestic use of drones toward the enhancement of the human race, however, has not been properly recognized, and continues to be highly unrecognized.

What might a drone operator do?

Elecmech Technician /Operator specialist

✖ Navigates Aircraft

✖ Will control air traffic flight patterns of drones before their departure through their destination.

✖ Maps designated and uninterrupted flight patterns allowed by FAA

✖ ($40.59 hourly)

✖ The ETOS will have specific training necessary qualifying them for their career position with */PS*. Their duties will include: navigating aircraft, controlling air traffic flight patterns before departure and through destination, and mapping of designated and uninterrupted flight patterns allowed by the FAA. After researching with the Bureau of Labor and Statistics (using mechanical and/or technical technician's pay scale compared to an aircraft pilot's wage) it should be appropriate that the ETOS will earn $40.59 per hour.

So, next on the agenda might be how to find new employees for your suggested new job market.

POSSIBLE INTERVIEW QUESTIONS /PS ETOS

✖ Structured Questions:

✖ This position may involve some travel, so are you willing to commute overnight?

✖ Drones may soon become a common place unmanned aerial system making package deliveries, as UPS (United Parcel Service) plans for today, for example. This involves some international correspondence on economic issues. Do you speak any language other than English?

✖ Do you have preference to indoor or outdoor work?

✖

✖ Situational Questions:

✖ Running late and behind your normal morning schedule, what would you do if you encountered an injured animal on your way to work?

✖ What would you do if you caught a co-worker stealing personal or company items?

✖ What would you do if you were witness to a drone (unmanned aerial system) flying overhead, unannounced?

✖

✖ Behavioral Questions:

✖ What has been the number one behavioral stress factor for you, while on a job (in class)?

✖ What has been a job behavior of which you have been exemplar, while remaining within acceptance of the norm?

✖ What would you qualify as a behavior or example in which you feel you have participated in a 'global' capacity (for any duration, in class, present or previous employer)?

Of course, performance evaluations will help your future organization solidify their 'people power.'

(Example: /PS Performance evaluation)

✖ *Based on the ETOS job, functions, and specifications, it is necessary that this evaluation be*

✖ *based on a combination of an employee's traits, behaviors, and (as well as) results:*

✖ Employee arrives to work *on time*.

✖ 1   2   3   4   5

✖ Employee has completed training for this quarter.

✖ 1   2   3   4   5

✖ Employee displays *initiative* while gathering GPS/Satellite, and statistical information.

✖ 1   2   3   4   5

✖ Employee has shown willingness to work overtime when requested.

✖ 1   2   3   4   5

✖ Employee displays *positive behavior* when working in team situations.

✖ 1   2   3   4   5

✖

✖

✖

✖

✖

✖

✖

✖

✖

✖

✖

- ✖
- ✖
- ✖
- ✖
- ✖
- ✖ Employee is reliable and *dependable* when job assignment warrants working independently.
- ✖   1   2   3   4   5
- ✖ Employee has worked one holiday this quarter.
- ✖   1   2   3   4   5
- ✖ Employee has shown work-home life styles with interest and some *enthusiasm*.
- ✖   1   2   3   4   5
- ✖ Employee offers specific suggestions for improved work environments.
- ✖   1   2   3   4   5
- ✖ Employee has passed physical wellness check for the quarter, showing dexterity for physical operation of equipment.
- ✖   1   2   3   4   5
- ✖ Employee *has demonstrated* the ability to implement necessary changes on short notice, for example when weather poses problem for UAS flight.

1   2   3   4   5

- ✖ Employee has been able to safely control air traffic flight of */PS* drone from departure through destination.

1   2   3   4   5

Due to the job, function, and specification described in the ETOS position, a performance evaluation will be based on employee's traits, behaviors, as well as determining if the ETOS has produced the required results when performing his/her job.

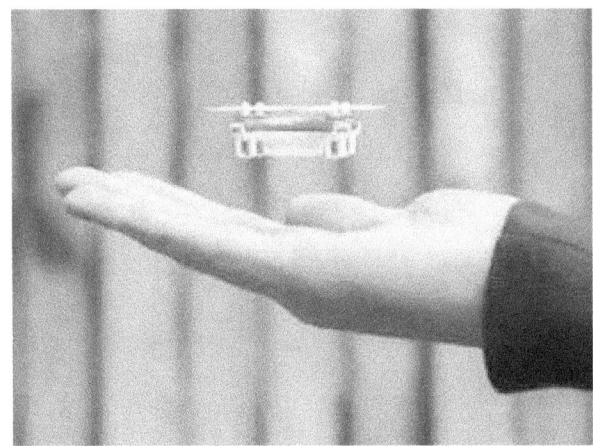

While creating your job of the future, don't forget to create your code!

Code of Ethics

/PERSONAL SPACE CODE OF ETHICS

✖

✖ /Personal Space will maintain safety and comfort while providing the use of its UAS.

✖ (Unmanned Aerial System, PS-DRONE) is utilized for purposes of inspection, severe weather conditions, and/or assessing police and fire issues.

✖

✖ An employee of /Personal Space must maintain U.S. citizenship. Illegal behavior, including discrimination and harassment, is not tolerated and will accrue additional punishment from this organization and its board members. Relationships with co-workers and/or clients require attention to detail, as well as the passion for the development of human welfare. All employees are expected to reflect their social responsibility with great care, and conflict of interests will hold liable those whose actions are presented. Political and community involvement will reflect those ideals set in equality and shared by stakeholder representation. Bribes are unacceptable.

✖

✖ The values of each employee inspire the creation of an ethical framework that will implement company reason, regulation, and review.   Stewardship of these responsibilities begets the financial stability which will secure future growth for /Personal Space and all of their employees. /Personal Space remains transparent in all of their accounts and documentation.

✖

✖ /Personal Space requires a statement of values and code of ethics from all employees, in particularly verbalizing that: **"As a member of the /Personal Space team, I will arrive for work, performing my tasks to the best of my ability on a daily basis. I will maintain a positive attitude in diverse situations, accepting what may seem as unacceptable. I will challenge what may be out of the ordinary by expecting from others that which I would only require of myself."** (A disclosure statement might follow.)

✖

Our code of ethics is reflected in each of our employees.   U. S. citizenship (a precondition for this position) does not guarantee that the employee is socially responsible, and each employee must work hard to develop an ethical framework which presents /PS as transparent in all business affairs.

Knowing what you want to do in your future job means knowing your own strengths, weaknesses, opportunities, and threats.

/PS drone operation might involve:

SWOT

- **Strengths**
- Has strength in market for delivering customer's requested project
- Diversity of options available/several UAS to choose from
- Weekly 'updates' from Federal Aviation Administration
- Market share in virtual training
- Capacity number of employees
- Many work locations
- Contracts completed per customer requests, ongoing
- Family / Friendly work life balance Day Care program *is* planned on site
- Elderly Care program *is* planned
- Employee Specifics Program and/or EAP *is* planned
- Career Advancement & Continuing Education
- Overtime/Work from home
- **Weaknesses:**
- Newer equipment (drone) is not recognized in civil environment
- Public unawareness information available
- Need for more Research and Development
- Nature of industry (commercial, domestic/civil) not defined
- Legislation still pending
- **Opportunities:**
- Emerging technologies
- New market use
- Global expansion
- Setting Trends
- Work with the Federal Aviation Administration
- **Threats:**
- Competition
- Unusual bird flocking
- Unknown civilian Arial apparatus and/or model airplanes
- New Regulations
- Customer's perception of */PS* drone in weather related and emergency situations
- Protests/Boycotts
-

✖ Each component of your company's strengths, weaknesses, opportunities, and threats should be considered.

... Now you start thinking of a job....

Have a strategy! Develop your own plan.

Strategic plan example for ETOS:
- ✖ *Strategic plan to be implemented:*
- ✖   */Personal Space* has strength in the market share of virtual training; we are

able to provide equal employment opportunity status on a daily basis. Relationships with

manufacturers, to acquire the current supplies required when assembling /PS drones, will

represent those companies with strategies of the same interest. With any potential risk of

an elevated ethical conflict situation, */Personal Space* will abide with all laws and

regulations affiliated with unmanned aerial systems. Compliance with government

and/or officials involved will minimize audit requirements. Our values in people and

planet will provide guidance in acquiring profit. Research and Development will be on-

going, tapping emerging technologies as a priority for expanding new markets. Media

announcements will be made when necessary, including the delivery of drones to areas

where weather-related and/or emergency situations arise. The implementation of Human

Resource Management will address employee care, child care, and elderly care to provide

the best working atmosphere possible. */PS* intends to remain competitive.

✖

✖   */Personal Space* hopes to address the public outcry and reform needed to make UAS (drones) applicable for the common good for all mankind, and in a profitable manner.

✖

✖   Being tested by companies such as Fed Ex and Amazon, 'hobbyists' are now hoping that their operations will also be allowed. */Personal Space* hopes to initiate the public understanding that is necessary so that civil uses for UAS (drones) remain in check until public understanding properly recognizes the potential such technology may pose for *invasion of privacy, human operational error, and offensive conduct.* Our hope, for the future, is that unmanned aerial systems are allowed.

✖

✖   */PS* will continue to work with any legislation which is offered, and hopes to qualify the <u>ElecMech Technician and Operator Specialist</u> (ETOS) with the training necessary for accomplishment in such a career position. Whether qualifying missions may be locating people in offshore oil-rig fires, pursuing cars that flee accidents, or spotting cattle escaped from farmland, the ETOS will be prepared in their aeronautical science. However, training will be subject to new official regulation in an on-going capacity; training will include 200 hours in simulator and hands-on experience **before** attempting any job assignment.

Finally, your new business plan can be presented allowing your suggested new job position to become an accomplished and successful career position for many who seek employment.

Develop *your* business plan for your newly suggested career position.

I have suggested a job of the future. What might you suggest?

Let's get started.

Additional research completed on this publication:

Management Component:

    A. Job/Organization Identification:

Drone ElecMech Technician and Operator Specialist

Organization's Name:

/Personal Space

Starting Hourly *Base*: $40.59, see detail @ blc.com

B. Job Statement:

    The ElecMech Technician and Operator Specialist controls the air traffic flight patterns of drones before their departure through their destination; maps designated and uninterrupted flight patterns.

C. Essential Functions:

    Holds 4-year degree in Bachelor's with aeronautics background, preferred

    Is familiar with satellite operations and programs GPS

    Navigates aircraft

    Works through the critical time span related to weather-related or emergency calls

    Determines if job request warrants assistance of drone

    Checks local community regulations before implementing use of drone

    Works in close communications with FAA in determining flight patterns available

D. Job Specifications:

    Passes standardized testing given by FAA on bi-weekly basis

    Follows FAA guidelines

    Maintains company protocol

    Complies with after training requirements

    Passes medical wellness check exam, yearly-for first three years, then biannually

    Works one holiday shift per year and weekends, if required

Our future jobs depends on our ideas and input.

Input yours.